Some farmers were totally converted to Perkins power as can be ascertained from this posed shot between round ricks with P3(TA) converted Fergie 20s, a Nuffield DM4 with P4(TA) and a Fordson Major with L4(TA). The savings, and convenience of an engine which lacked the temperament of the TVO units then usual in tractors soon offset the initial extra costs.

PREFACE

It is now ten years since the first printing of this particular book and continued interest has made it possible to provide a reprint, which apart from this revised preface and a new cover, is in all respects identical to the first.

The debate regarding diesel engined tractors at Vintage Ploughing Matches continues. The first printing of this book did much to ensure that owners of genuine conversions were not penalised when entering and attending these events.

What is more important however, is the fact that in addition to the first printing of this volume, the "Vintage Tractor Special" and "Classic Tractor Special" series continue to be some of the best selling books on the subject ever published. Where appropriate the most popular titles almost never go out of print and the slower sellers such as this edition have every opportunity of being available again at some time in the future.

With regard to the Perkins story, that engine builder is very much still alive today; but the pioneering years and the impact which Frank Perkins' interest in diesel conversions for tractors had in the 1940s are an important part of the story of mechanised agriculture in the 20th century.

Allan T. Condie.   Carlton.   April 2000.

PERKINS DIESELS

The diesel engine had been around for some time in various forms, indeed its application in heavy goods vehicles and PSVs was becoming almost universal by 1945. Various attempts had been made to launch diesel tractors onto the UK market, but most of these had fallen by the wayside. Strange to say there was not one totally new multi cylindered diesel on the British Tractor Scene pre-war. The AGE tractors using Blackstone and Aveling engines were basically International 22/36 frames with diesel engines fitted. A few International Petrol Start Diesels as fitted to their WD40 and TD35/40 Crawlers were imported, but like the domestic conversions they did not amount to much. Mainstay of the Compression Ignition picture was the Marshall, but even there production in total hardly got into four figures. As the majority of UK tractors were in any case multi cylinder units with automotive type engine and transmission layout, it was obvious that any manufacturer setting out to build tractor engines would adopt the multi-cylinder concept.

One of the immediate problems which arose in producing a diesel engine was its cost, not only in development terms, but in production also. Much finer tolerances were necessary and stronger components desirable to withstand the much greater compression pressures of such engines.

Barford & Perkins of Peterborough became part of Agricultural and General Engineers (AGE). Their main product line was motor road rolling equipment. The Peterborough works was closed by AGE when they took over in 1928/29 and their road-roller business was merged with that of Aveling & Porter to form Aveling-Barford.

Mr Frank Perkins moved to Aveling's works at Strood as Works Manager. It was while he was there that the Vixen engine was designed, the drawings being done by a Short Bros. (Seaplanes) draughtsman working at weekends in the cellar of Mr Perkins' house. One or two engines were built at Aveling's before the big decision was made to form a Company to produce them.

Frank Perkins Ltd. was formed and premises at Queen Street, Peterborough, formerly occupied by Barford & Perkins, were rented. This was a case of 'coming home'; Frank Perkins obviously knew that the works were standing empty, and an approach to the owners, Milton Estates, secured a lease. Indeed, quite a lot of equipment, such as the benches, were still there. Here the application of medium sized diesels for industrial and marine work in the nineteen thirties, was pioneered. The Leopard, Wolf and Lynx engines were basically intended for multi use applications, covering vehicle, industrial and marine installations. However the use of the Leopard engine in tractors was instigated in the mid thirties. A Fordson N Land Utility tractor on Firestone wheels and tyres was supplied to F. Perkins Ltd. on 11th February 1937 by Willenhall, Staffs, Fordson dealer Reginald Tildesley. This was serial number 808499 and it was delivered to Peterborough less engine.

A Perkins Leopard engine serial number 7343, built on 16th June 1937 was fitted. This was a Leopard II engine with cast iron pistons and was rated at 34bhp at 1100 rpm, it being derated from the usual industrial rating of 46bhp @ 1500rpm to prevent damage to the rear axle.

The prototype tractor was tried out at Tettenhall near Wolverhampton in July 1937, prior to its exhibition at the Royal Show, which in 1937 was at Wrottesley Park nearby.

In October and November 1937 a further 11 engineless tractors were supplied to Perkins ex Tildesley, and in 1938/9 a further 17 were supplied, making 29 in all. Incidentally, the TVO engines were used to provide a 'float' for Tildesleys' exchange engine scheme on the Fordson. The first Leopard conversion to go into full time farm service was bought by Mr T.R.C. Blofeld of Hoverton Fruit Farms, Wroxham, Norfolk. It had the reputation of being a bad starter, but the early models had no self starter and were started on the handle using a decompressor. It would appear that most of the others were exported, and there is at least one extant in Australia and one in New Zealand. The latter example has a self starter. Another engine was also installed in a Muir Hill 3 cu yd. twin wheel dumper which appeared at the Public Works Exhibition at Olympia in 1938 - this also had electric start.

With the advent of the war, conversion of tractors ceased, but in the meantime the development of a new range of engines was under way at Peterborough.

The result was the P series of engines which came onto the market in the early war years. By taking the design from the outset to produce three, four and six cylindered units using common parts saved much in development and production costs. Incidentally, one and two cylinder prototypes were built but never produced in any quantity, and it took some time before the three cylinder variant appeared. In fact, one P3 prototype was built in 1939, and fitted in a London Taxi, but it was to be 1951/2 before the engine was developed. The P series originally had names as had the other pre-war engines - the P6 was the Panther, the P4 the Puma, and the P3 the Python. In fact, the P4 was introduced November 1937 and the P6 in February 1938. There were of course two versions of the P series. That for use in applications where engine speeds of over 1500RPM were required had aluminium alloy pistons, and that for use where engine speeds of below 1500RPM were required had cast iron pistons. The calibration of the fuel injection pump produced engines with very different torque characteristics. There were common parts amongst all 3,4 and 6 cylinder variants, all of which shared the same bore and stroke of 3.5" x 5". The swept volumes were P3 - 2.36 litres; P4 - 3.14 litres, and P6 - 4.73 litres.

The design of the engines was such that different sumps, flywheel housings, front and side mounting plates, and exhaust/inlet manifolds allowed a diversity of use, whether with a wet clutch application such as the Fordson Major, or a dry one such as the Massey Harris 744D. For vehicle use of course mountings were different, but the sump used with Industrial applications had sufficient strength, when necessary, to include the engines in 'unit' applications. There was even a petrol version of the P6, developed in conjunction with Dennis Brothers of Guildford, to replace their equivalent petrol engine; two were built, and were only experimental.

The P series had not been used widely in tractors until Frank Perkins converted a Fordson Major for his own use. The result was that Ford Motor Co. sent two Majors for conversion, one with a P4, and the other a P6. Both had fabricated sumps and flywheel housings. Whilst both gave a satisfactory performance, the P6 was chosen due to its lower maximum speed which suited the E27N gearbox better, and in adopting the engine as a production option it also must be remembered that the Ford dealer network were already geared up for Perkins spares and service as the engine in its vehicle (V) form was fitted in the Thames lorry.

So it came to pass that the P6(TA) engine was born and graced a good few thousand E27Ns, either straight off the production line or as conversion packs which Perkins sold themselves.

Morris Motors were also looking for a diesel to fit in their new Nuffield Universal tractor. They chose the P4(TA) which fitted the Nuffield chassis in more comfortable form than the P6 fitted the E27N. In the case of the Fordson a special sump had to be cast to support both front axle and connect this with the flywheel housing. Whilst most P series engines were made for dry clutches, the E27N's wet clutch also required a special flywheel housing and suitable oil seal for the starter motor.

As we shall see in due course, other manufacturers made use of the Perkins engines in their (TA) Tractor adapted form.

One feature of the adoption of Perkins units in some applications was the fact that the power output of the engine was much more than the equivalent Petrol or TVO models. Also the initial cost of the diesel engined version was much more than the equivalent spark ignition one.

With Ford and Nuffield offering Perkins engined variants, and David Brown their own engine, it was not long before Ferguson's sales force were calling for a diesel version of the TE-20. Now as Harry Ferguson himself was not a diesel fan, it took some considerable persuasion to get him to agree to a diesel engined tractor at all.

The Perkins P3 was looked at, but the cost of installation and the modifications needed would not suit, so a Standard engine was fitted instead. This engine was designed by Freeman-Sanders specifically for Ferguson, and was built for them by the Standard Motor Co., who later developed the same unit for use in their 'Vanguard' car.

There was one thing which opened the way for a conversion pack from Perkins for the Fergie 20 and that was the fact that the diesel engine supplied by Ferguson could not be fitted to existing tractors. The first tractor converted by Perkins was actually a Ford-Ferguson 9NAN for his own use, but the principles of the conversion were the same for the TE-20. The 9NAN was acquired by Frank Perkins during the War and the engine fitted in this case was a P4, there being no P3's at that time. It was cut up in the mid 1970s under Customs supervision, along with an MF85 imported as a test tractor for the A4-300 engine, as no import duty had been paid on them!

Massey Harris were looking at building tractors in the UK, as there was great export potential, and the government were supporting the 'ground nut' scheme in Africa. A few type 44 frames were taken and fitted with the P6(TA) engine. Tractor assembly had commenced at Manchester in 1948. Due to shortage of space, the 744 as it became was then assembled at Kilmarnock from 1949.

Thus, not only were Perkins engines now being fitted in production by leading manufacturers, they were also available to convert s.i. engined units to diesel. By the late forties 'conversion packs' were available to convert the lions share of popular models.

The promotion of Diesel conversion packs was just as well, as new engine business fell off considerably with Ford fitting their own diesel to the New Fordson Major, and Nuffield fitting the BMC engine to the Universal by the early fifties.

The P series was not without its problems in agricultural service. Unlike many road hauliers and others who used the (V) "vehicle" version, farmers were not renowned for their care and attention to machinery, and a diesel engine requires more of this to maintain peak performance. The chain driven timing could, if not properly adjusted, cause starting problems, and early tractor applications had engine breather problems later cured by modification. The tendancy to over-rev. the P series (TA) units with their cast iron pistons often caused bent con-rods, little end failures, and cylinder head gasket failures. Indeed later P6 engines all had alloy pistons which alleviated many of the problems, and many rebuilds received these also. This is why a lot of P6(TA) engines still extant appear to the unknowing to be "V" or vehicle ones! The main problem with the P6(TA) was the farmer's habit of leaving the engines running when not working, as they were in the habit of doing with paraffin engines. This caused the exhaust ports to choke with carbon and Dereck Lambe tells me that he has taken off manifolds to see ports with holes not large enough to push a pencil through. Of all the tractors fitted with the P6(TA) the Fordson Major E27N was the most successful. Early models had the habit of breaking half-shafts; however a new design with suitable heat treatment cured this, and the E27N back end was really the only one which could stand the full power of the P6, and indeed this was the tractor which has endeared the P6(TA) to present day collectors and enthusiasts above all other. Of course more were built than any other diesel engined tractor of the era.

1953 saw the introduction of a completely new engine for industrial use - the L4. This unit was designed expressly for low speeds up to 2000rpm. It had a bore and stroke of 4.25 x 4.75", a swept volume of 4.42 litres, and could be set to give up to 59bhp @ 2000rpm. A gear driven timing arrangement was fitted and the camshaft, unlike the P series, was in the normal position with push rod operation of the valves. The cylinder liners were, however, of the wet type.

The new engine was adopted by various manufacturers for use in tractors. The Massey Harris 745 used the L4(TA) from 1953, and M-H also adopted it for their combines. Just to confuse the issue the development tractors were 744 with the new L4 engine fitted! It was also available as a conversion pack for Old Fordson Major, New Fordson Major, Nuffield, and International Farmall M tractors. Later it was adopted as the power behind the new Track Marshall crawler.

Perkins dealers were keen to seize any opportunity to convert tractors to diesel, and indeed a few 'one offs' were built to satisfy customer requirements. Where a popular model was involved it was possible to supply the complete 'conversion pack' to suit. The approved and popular conversions were:-

P3(TA): Ford 9N/8N. Ferguson TE-20. Farmall H. Allis B.
P4(TA) Nuffield Universal.
P6(TA) Fordson Major E27N.
L4(TA) Fordson Major E27N. New Fordson Major. Farmall M. International T6. Minneapolis Moline UT. Nuffield Universal.

It is believed that the most popular International L4 conversion was for the T6 crawler.

Perkins engines were also fitted into limited numbers of International W4, W6, and W9 tractors, A few Oliver 80s and 90s, a few Massey Harris 203 and 55 models, and a few Case D series. There was no approved conversion of the Fordson 'N'. It was naturally possible to fit either a P6 or a L4 to the standard Fordson, but as part of the agreement with Ford Motor Company which allowed Perkins to sell conversion packs to Ford dealers, it was agreed that conversion of the 'N' would not be encouraged. In any case the rear axle was considered not to be capable of taking the increased torque of the diesel. The agreement also allowed for supply of certain parts for the conversion packs from Ford's parts operation. Ford did not want the life of the thousands of 'N' models to be extended by dieselisation.

For the next chapter in the story we move back to the events taking place at the Ford Motor Company at Dagenham. Fords had long wished to build a model in the UK to compete with the "grey menace" or Fergie 20. It was quite possible to do this, but the cost of developing an engine for such at tractor, which by the mid fifties had to be a diesel, was prohibitive.

The Ford boys knew of the original Perkins conversion of the Ford 9NAN, and as the new tractor was very much an updated 8N Ford's took some P3 engines for evaluation during development of the Dexta. These were however a development version which had a simple gear driven train and a small Holburn-Eaton type oil pump driven from underneath the crankshaft timing gear. This allowed the CAV in-line injection pump to be flange mounted directly onto the back of the timing case. All external oil and water pipes were eliminated. It was called the P3.144 and eliminated the expensive features of the P3, the skew driven oil pump, and the chain driven timing. A four cylinder (P4.192) variant was also built. Although the P6-288 had a DPA rotary fuel pump it retained the chain timing. The gear timed version was built in France as the 6PF, and was for Industrial use only (no vehicle version) and was never available in the U.K.

At Ford's request, Perkins fitted their 3 cylinder engines with Simms Pneumatically governed fuel injection pumps and injectors and also removed the company's name and trade marks from all parts. Perkins also designed the dress items such as exhaust manifold, water outlet housing, and gearbox adaptor plate. The engine went into production as the F3 in August 1957 and all were built at Peterborough.

In the early 'sixties the cylinder bores of the P3/144 were opened up to 3.6" diameter in common with other engines in the P series. A C.A.V. DPA fuel injection pump was fitted and the engine became the 3.152. The F3 was also uprated and a mechanically governed 'Minimec' version of the Simms in-line fuel injection pump fitted - this was known as the F.3/152 and was supplied for the Super Dexta from February 1962. Again all were built at Peterborough.

The engines supplied to Ford had all the blocks and heads cast at the Dagenham foundry of FMC, the rough castings being supplied to Perkins on a 'free issue' basis. The logistics of this operation were remarkable. No large stocks of engines were held, each day Perkins built just enough engines to cover the Dagenham assembly tracks requirement for the next day. These engines were delivered by Perkins' own lorries overnight. A small buffer stock of one day's supply of engines was kept in a special building at Peterborough to cover for any breakdown on Perkins' assembly line, and a similar number was held at Dagenham in case of breakdown or accident to any of the lorries on the delivery run. These lorries brought back the rough block and head castings as a return load, and the whole system seems to have worked remarkably well for the seven years that the contract lasted. This system is known today as 'just in time'. The Japanese fondly believe they were the first to introduce it!

Some 153,322 F3, and 64,496 F3.152 engines were supplied to Ford, a grand total of 217,818 engines between August 1957 and October 1964.

Perkins became part of Massey Ferguson Limited in 1959, and this consolidated the use of Perkins products in MF tractors. In fact from 31st August 1959 Massey Ferguson had taken over the Banner Lane Coventry plant of the Standard Motor Co. This did not involve any engine manufacture, and M-F were keen to move to a stage where their new subsidiary would provide all the engines for the tractor plant. The Standard 23C engine had a reputation as a bad starter, especially in cold weather.

The first Perkins engines used in a Massey Ferguson (as opposed to a Massey Harris) tractor were fitted in the new M-F 65 tractor introduced in 1958. The Perkins 4.192 engine was chosen to power this tractor and this gave the UK market a 50HP plus tractor with the Ferguson System which put the competition in the shade somewhat. It is interesting to note that the forerunner of this tractor, Frank Perkins' own 9NAN with P4 fitted existed some ten years earlier. Also, one of Harry Ferguson's LTX prototypes was fitted with an L4 engine. It took M-F that long to shake off the 'light tractor' ideals of Harry Ferguson, who was not in favour of high powered tractors, but believed in the lowest powered engine that would do the job with his draft control system.

Shortly after the M-F control of Banner Lane took effect, the M-F 35 was given a three cylinder Perkins 3.152 engine.

In 1961, a direct injection version of the 4.192 engine, which in its bored out form to 3.6" was the 4.203, came into production, and this was known as the AD-4.203. This was used from its inception, in the M-F 65.

With the Fordson Super Dexta on the market, ironically with engines from Perkins, the 35 was given the A3.152 engine from 1962. Perkins also supplied A3.152 engines for Ferguson's Detroit plant which were fitted to MF35s there, and also MF50 tractors. The Dieselmatic 65 used the 4A.203 engine in the USA, and later the AD4.203, whilst the Super 90, only built in the Western Hemisphere used the A4.300. This engine was a bit more than just a bored out 4-270. It had a dedicated tractor block, with cast-in tractor fittings, dry type cylinder liners, a 5 bearing crankshaft, and a harmonic balancer unit.

The 4.270 was an update on the old L4 engine, and brought direct injection to this engine, plus the use of a distributor type fuel injection pump. Conversion packs for the older tractors now used this engine instead of the L4, and other manufacturers such as Marshall took the newer engine instead.

Allis Chalmers had fitted Perkins P3 engines in their model B at the Totton Southampton plant, and continued to use Perkins engines - the P3.144 in their D270 and D272 models once assembly had moved to Essendine. The ED40 tractor used a Standard Motor Co. 23c engine however. This was a tactic used by Standard to annoy Perkins and M-F by selling engines at lower prices to take up the loss of production encountered with the loss of M-F business. They also got in to Fords by supplying a limited number of Petrol 87mm engines for Dextas sold in Denmark and elsewhere. These latter were, however, of special build to suit the Dexta gearbox housing.

Small numbers of engines were supplied to other manufacturers both at home and abroad, and examples of some of these are shown in the book. Notable crawlers which used Perkins engines were the Howard Platypus, the Bristol, and the French built Continental.

There we leave Perkins for the moment. As the sixties progressed and the seventies dawned an even greater variety of products came out of Peterborough, and a new factory was built. But that is another story!

(Full details of Perkins engines fitted to Massey Ferguson tractors from the late fifties until the mid seventies will be given in a forthcoming new book on Massey Harris, Ferguson, and Massey Ferguson Tractors).

The original Leopard conversion in the Fordson N taken at the Royal Show at Wrottesley Park near Wolverhampton. This tractor was number 808499. Note the Firestone rear 24" wheels in this illustration.

The 'Leopard' conversion being demonstrated at Tettenhall prior to the Royal Show in 1937. Personnel from Reginald Tildesley are: Bert Brandon - driving; Sidney Sharpley (Salesman) leaning on tank, then Reginald Tildesley himself. Perkins personnel are Mr G. R. Guest on the extreme left, in shirt sleeves, who was the Industrial engine salesman, Mr C. Kent, Service Engineer on the extreme right in the white smock, and 2nd from the right in the dark jacket Mr L. W. J. Hancock the Sales Manager. The person third from the right in shirt sleeves is believed to be a Mr Ellis who was with Perkins as an agricultural adviser for a short time in 1937. Cyril Kent, the last survivor of these people, died in the Autumn of 1989, aged 84, so there is now no-one to tell us about these jobs.

The other side of the tractor, fitted with spade lugs for demonstration. It had the green spot transmission. Starting was achieved by decompression and the use of the starting handle.

The Muir Hill 3 cu.yd. twin wheel dumper on display at the 1938 Public Works exhibition at Olympia. This unit was an amalgam of Perkins, Fordson, and Muir Hill's own parts, but did feature a self-starter which can be clearly seen under the exhaust pipe.

Two shots of the 'Leopard' Fordson taken by an unknown photographer at the same demonstration as opposite. The view above shows clearly the exhaust manifold arrangements, the oil filter, fuel injection pump, while the view to the right shows the water pump mounting at the lower right hand side of the engine, plus the way in which the upper dash is mounted higher and further forward than normal to provide for the increased height of the engine, which was rated at 34BHP @ 1100rpm.

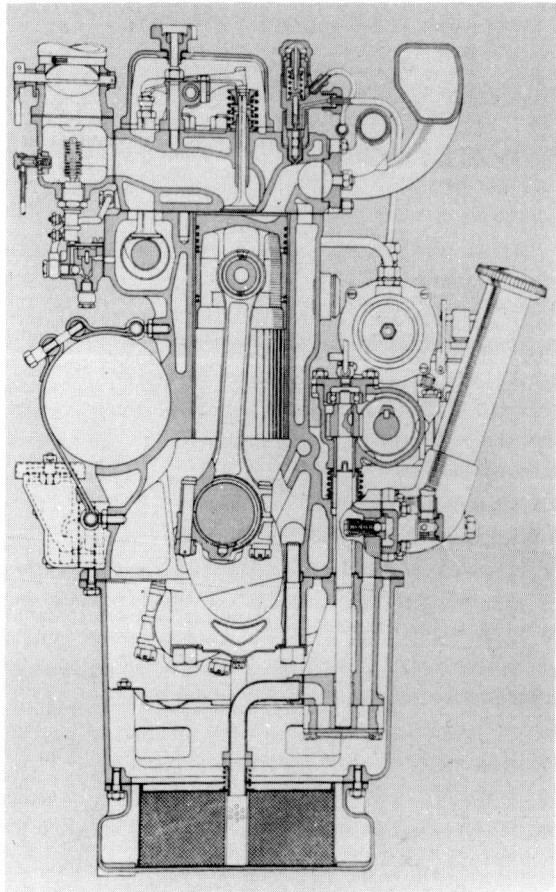

A shot of what is believed to be the original P6 engined Fordson Major E27N in the works yard in March 1947. A P6(I) engine was used, and some evidence of the use of existing parts can be seen at the front of the engine sump. Note the location for the battery, above the right hand half-shaft housing. The original engine in this tractor threw 'a leg out of bed' whilst being used to haul sugar beet. The driver complained about its non performance whereupon 'the boss' (Frank Perkins) told him to open her up! A wide open throttle - and BANG! Note the Perkins standard starting handle with forged crank and brass hand grip. Later engines used the standard Fordson product modified.

Left: The P6 engine in section.

Opposite page Top: The Prototype supplied to Ford with P4 engine. The neatness of the conversion is to be remarked upon. Note the use of existing electrical arrangements. The first few conversions had a 90 degree adaptor fitted to the induction manifold. This required the pipe connecting it to the air-cleaner to have a double bend in it, as on the P4 shown here. It was expensive and quickly modified.

Opposite page Bottom: The prototype supplied to Ford with P6 engine. An early tractor must have been supplied for conversion, judging by the radiator. A close look on the steering column will reveal a made up panel for the engine stop control, oil pressure gauge and the Kigass pump. This tractor has the adaptor at a little less than 90 degrees to the induction manifold, allowing the connecting pipe to be straight.

Opposite Page: The two prototypes in close-up, showing the P6 engine (top) and the P4 (bottom). All sump parts and flywheel housings were fabricated.

Above: The first production version of the P6(TA) for the E27N. The front mounting bracket required the use of a different radiator bottom tank. The oil filler on early versions was on the right hand side of the flywheel housing. The E27N version was rated at 46BHP @ 1500rpm.

Below: An early production version of the Fordson Major showing the grille with narrow bars adopted on the launch of the diesel to allow for better cooling. At this early stage, the existing primary air cleaner was used. This caused so much restriction that the engine's pneumatic governor came in early and maximum revs. could not be obtained. This led to the design of the domed top with angled vanes and 'rock slot' — a true centrifugal pre-cleaner.

Experience with the P6(TA) for the E27N showed up several weaknesses in design and these were rectified. This later engine, which was the final version, shows the revised combined breather/oil filler arrangements, and a larger fuel filter mounted on the engine. The top water connection has also been altered.

Below. The engine installed. Note the additional fuel filter and sediment bulb added in the fuel system. Farmers in those days were not prone to keeping diesel very clean! Note also the twin six volt batteries adopted in production and as supply in conversion packs.

Right. The right hand side of a later engine showing that the engine breather pipe has disappeared. On early tractors this used to allow the oil to run out when the tractor was running downhill on steep slopes. Note the location of the serial plate.

Below: The P6 in an E27N always looked impressive and fitted as if it were made for the job! Perkins always took pains to ensure that the installation was as neat as possible.

An impressive line-up of Perkins P6 power.

More Perkins power on display. This shot was taken on the premises of G. F. Slight Jnr., Hillside Farm, Cheapside Brigsby, Grimsby, Lincs. There were more County crawlers with P6 engines than TVO ones. The two Fergies and the wheeled E27N are obviously all conversions.

Opposite page top: With the arrival of the L4(TA) it was soon made available to convert existing E27Ns to diesel, but was of course never fitted in production. The front mounting was adopted to take the usual Fordson radiator without alteration.

Opposite page lower: The L4 installed. Note the reversal of the fuel tank, and the use of different batteries to those on the P6. The large dome shaped pre-cleaner, adopted early on for P6 engined Majors, is also seen here. The L4 in the E27N was rated at 45BHP @ 1500rpm.

A superb photograph of a L4 engined Major at work.

Below: The L4 with dress items to suit the New Fordson Major. This engine gave over 10HP more than the original Ford unit which was rated at 38HP, although a number of conversions were made to the E1ADKN (TVO) model which in its original form was not a success. A fuel lift pump is provided as with the Ford engines.

Two shots of a New Fordson Major with L4(TA) taken in Milton Street, Peterborough, outside the works. Note the alterations to the bonnet to fit two batteries, and the relocation of the badges. The fuel tank with its two fillers tells us that this was a TVO tractor (E1ADKN) originally, and the horizontal exhaust has been retained. This tractor was the prototype for the L4 conversion and is being driven by Vic. Corney, a fitter in the experimental shop who did the conversion.

A County Crawler with L4(TA) engine fitted. This tractor belonged to J.W.E. Banks Ltd., St. Guthlac's Lodge, Crowland, Lincs, who was an enthusiastic Perkins user.

Another E1A Major with L4 conversion, ploughing. This one has the vertical exhaust of an early TVO model. With the advent of the Mark II Ford engine in 1957, and the general availability of these Ford diesels, Perkins conversion packs fared rather badly on this model from then on.

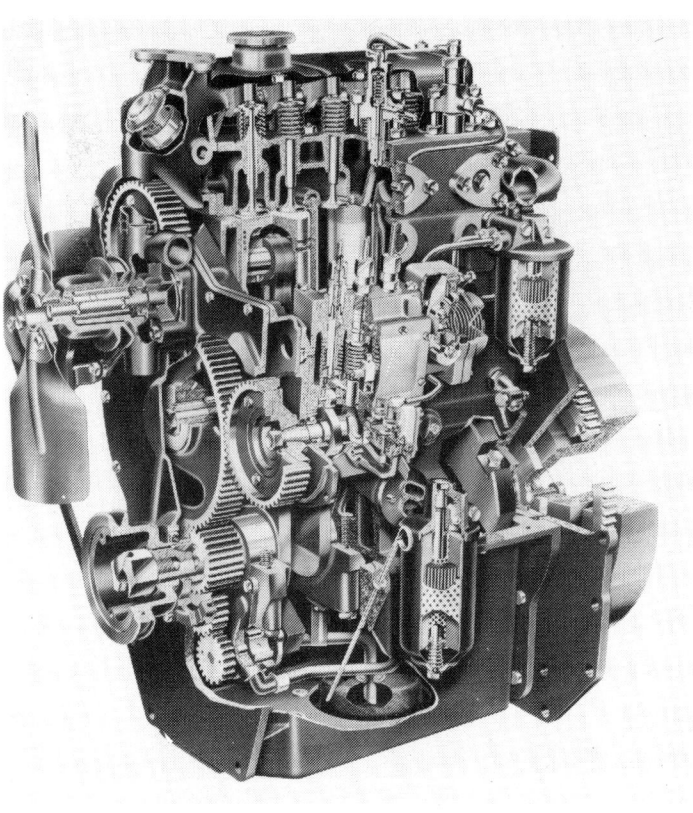

The F3 engine made for Ford for their Fordson Dexta is seen here in section. The gear driven timing, the oil pump drive, and the general tidying up of the design had its roots in the P3.144 (see page 23). The Simms injection pump was fitted for Ford use of course. The order for the F3 engines more than compensated for the loss of 'conversion pack' business on E1A (New Major) models.

The Fordson Dexta at work. This little tractor was one of the most successful of its day, and gave Fords tractor division a weapon to fight the 'Grey Menace' as Ford's salesmen called the Fergie. For its ancestry, see page 20. The original Dexta F3.144 engines had 3.5" bore and 5" stroke, but the F3.152 had the bore opened up to 3.6" when fitted in the Super Dexta.

Frank Perkins had a Mark 1 version of the P4 fitted into a Ford 9NAN tractor during the war. The above view shows the complete tractor with plough.

The shot below shows the right side of the engine, which in its Mark 1 version had the inlet manifold cast inside the cylinder head. Two very large 6 volt batteries in series were fitted; 'F.P.' made sure his tractor would start! The whole ensemble was somewhat longer of course, and the abandoned attempt to lengthen the front axle radius arm can be seen. The final solution was to weld steel blocks onto the fork ends of the radius arms and redrill the holes for the bolts through the axle beam. The flywheel housing was fabricated, and the use of an ordinary vehicle sump required the use of angle-iron braces between the flywheel housing and front saddle bracket. Note the huge Simms heater-starter switch on the left hand side of the steering column.

The engine from the left side is seen above. There is a dent in the oil filler pipe to give clearance to the steering arm on full lock - by accident or design? The standard P series water pump was used, and its high position meant using the radiator, fuel tank, and bonnet some eight inches. the fuel tank was also reversed, and the filler cap now extends through the bonnet. The pictures were taken in 1952 when the P3 was being developed for the Ferguson; the 9N had been in service for around 7-8 years.

The original conversion for the Ferguson is seen here. The engine was an Engineering Dept. prototype using standard Mark III P-series parts, which included the high position water pump, large oil filter and P4 Vehicle exhaust manifold. Note that the figure '4' has been ground out of the firing order.

Another customer receives delivery of a Fergie 20 which has been converted using a P3(TA) conversion pack. The Dodge lorry is a 1954 Dodge 'Kew' 7 tonner. It had the same Briggs Motor Bodies cab as the Leyland Comet and Ford Thames ET6, and was P6(V) powered.

A rather unusual Fergie with P3(TA) engine is this "Tracpac" crawler converted by a Leeds firm. The prototype with its high bonnet line encouraged the design of a smaller and much cheaper water pump to fit onto the front of the timing case cover, which reduced by half the amount that the bonnet etc. had to be raised. Design of a special, much smaller exhaust manifold and a smaller oil filter helped to keep costs down. The single large 1 litre size C.A.V. fuel filter of the prototype was replaced by a half litre C.A.V. and a Tecalamit pre-filter in series. The farmers habit of filling up his tractors from dirty old cans had been noted!

The P3.144 engine was a development of the P3 designed to reduce costs, and this example is fitted out for use as a conversion pack for the Fergie 20.

A TE-20 fitted with a 4.99(TA) engine. Only the one was ever converted, and its performance was lacking as the final-drive ratio was far too high. It did however form the test-bed for the French built MF 25 which came later.

The P4(TA) as equipped for fitting in the Nuffield Universal. A fuel lift pump was provided due to the distance of the tank from the injection pump.

A show picture of the Nuffield Diesel. Other interesting vehicles just visible are the JNSN lorry in the background, the 744 behind, hiding a Perkins engined road roller.

The P4(TA) fitted neatly into the Nuffield, and only required the relocation of the aircleaner.

The maximum rated horsepower of the P4 was 43BHP @ 2000rpm. This would enable the Nuffield here to easily cope with the scuffler in tow. The cab is by Scottish Aviation.

The L4(TA) was available as a conversion pack for existing Nuffield M3 and M4, PM3 and PM4 models. The view to the right shows it with all parts necessary to drop into the Nuffield frame.

An L4 engined Nuffield ploughing. Running at the maximum speed permissible for the L4, 2000rpm, which was also the maximum designed speed of the original engine, the L4(TA) would develop 59BHP.

Two more views of the L4 engined Nuffield at work. The axle extensions were a menace when negotiating narrow gateways, and many posts were knocked out this way. The aircleaner on the conversion has been relocated as shown, being originally on the opposite side of the tractor.

Above: An early 744D at work. Note in particular the rearward exhaust position, the wide cutaway of the bonnet top, and the wing mounted headlamps.

Opposite page: Two illustrations of the first Massey Harris 44 equipped with P6 engine as a prototype for the 744. Note the Dunlop pattern rear wheel centres, and the fitting of a Perkins badge - production tractors did not advertise the origin of their engines, although Perkins themselves always fitted a badge for publicity shots.

P6(TA) installation for a late Massey Harris 744D. The exhaust manifold has now been redesigned to take a vertical silencer, and breather/oil filler modifications made. The engine in the 744 was rated at 46BHP @ 1350rpm. Paper fuel filters fitted as part of the engine build are now evident.

Above: A late 744D showing the short wings, improved battery boxes, and later type engine.

Below: The L4(TA) engine as turned out for the Massey Harris 745.

Opposite page top: The development 744 now fitted with a L4(TA) engine as the 745D prototype. The use of an earlier bonnet belies this - very early and late 744D tractors had a much greater cut away. The L4(TA) in the 745 gave 50BHP @ 1500rpm.

The photograph may have been taken at Racine, Wisconsin, USA; as the new engine was tested there before being adopted for production. It is said that the idea for the L4 was developed from the Continental HD260 Diesel removed from one of the 44s sent to Peterborough by Massey Harris to be fitted with a P6(TA) as a development tractor for the 744D.

Opposite page lower: A 'one off' just to show that there were such things in the old days. This Massey Harris 102 Junior has been fitted with a P3(TA) by Perkins' agents in Rhodesia.

The P3(TA) as dressed for installation in the International Farmall 'H'.

Below. Some neat design work is needed to keep the original bonnet line of the Farmall 'H' intact. Note also the battery mountings.

Opposite page top: The complete Farmall 'H' with P3(TA) is seen in the upper picture. Note the front wheel weights.

Opposite page bottom: Another one off. This McCormick Deering W6 has been fitted with a P6(TA) engine.

This page: Two views of a Farmall M fitted with an L4(TA) diesel engine. There were quite a number of these tractors converted with such engines.

Opposite page top: One of the most popular L4 conversions, seen here hidden from sight, was to the International T6 crawler. A few W6 wheeled tractors were also converted, as the conversion parts would suit.

Opposite page bottom: The International 300 was only produced for just under two years in the USA. It was IH's answer to the Ford NAA and Ferguson TO35, but failed to meet with the success expected. Here is one fitted experimentally with a Perkins L4(TA) engine.

Allis Chalmers adopted the P3(TA) as the diesel power unit for its model 'B' tractor assembled at Totton near Southampton. Conversion packs were also sold to adapt both British and US built B's to diesel.

The installation of the P3(TA) into the Allis 'B' was perhaps the least tidy of all the conversions due to there being a lack of space to put things on this wasp like machine.

Opposite page upper. Use of Perkins diesels continued through the D270 which was very similar to the British built Allis B, to the D272 seen here, which used the P3.144 engine.

Below: Whilst Allis Chalmers in the UK went to Standard for the engines in its last tractor, the ED40, the French built FD3 retained the option of a Perkins P3.144 engine.

The L4(TA) was also available to fit the MM UTS, and it is seen (right) with all necessary parts to fit.

The left side of the installation in the MM UTS.

The right hand side of the installation.

Two views of the MM UTS with L4(TA) engine fitted. Tractor mythology has it that this conversion was done initially at the instigation of Sale Tilney, MM importers in the UK, to complete some UTs tractors which had been salvaged from a ship sunk in Liverpool Docks during the war, whose engines were no use.

The Marshall organisation started to fit Perkins L4(TA) engines as seen here to their Track Marshall crawlers in 1957.

Many of the Track Marshalls built in the 1950s and 1960s are still at work today, and Marshall's still use Perkins engines. This shot dates from the late fifties however, and shows one of the early models at work.

*Even the mighty Cat D2 was not immune from being fitted with the Perkins L4(TA) engine. This example was operated by J.W.E. Banks, who also owned the County Crawler shown on page 18.
The fitting of the Perkins engine would eliminate the need for the starting donkey engine of the CAT diesel.*

The little Bristol 22 crawler made use of the Perkins P3(TA) engine.

The Howard Platypus was P4(TA) powered and is seen here in 'narrow' form, above on show, and below working in the hop-fields.

To finish the section on conversions we illustrate this further 'one off' - a L4(TA) engine fitted in a Case DC4.

Cockshutt offered their model 40 de-luxe with a Perkins L4 diesel in the mid fifties.

Once Perkins had been taken into the Massey Ferguson fold the adoption of the Perkins P3.152 engine for the 35 was a natural progression. The top illustration shows a 35 so equipped and the two pictures above show the engine, which was of 3.6 bore and 5" stroke giving 35HP

The MF65 used the Perkins 4.192 engine to give 50Hp in a British built Ferguson tractor for the first time. Later tractors used a direct injection version of the bored out variant, the AD4.203 which gave 58.38BHP @ 2000rpm.

The 35X also used the bored out version of the P3.144, the A3.152, built from 1962.

Above: The MF65 was Massey Ferguson's answer to the Fordson Major Diesel and Nuffield Universal Four tractors, both of which started life with Perkins diesels of course. The US built version is seen with different tinwork (below).

The MF50 was not available in the UK but used the Perkins 3.152 engine (right).

Conversion pack business continued into the early nineteen-sixties and here we illustrate a selection of the then available engines. Firstly (left) we have the 4.99(TA) which had a 3" bore and 3.5" stroke and whose maximum rating was 35BHP @ 3000rpm.

The P3.144(TA) replaced the P3. It had the same cubic capacity, but gear driven timing, and was rated at 35BHP @ 2000 rpm.

The 4.192(TA) replaced the P4. It also had the same cubic capacity as the P4, and had a rating of 50BHP @ 2000rpm, and was fitted with a DPA fuel injection pump.

The 4.270(TA) replaced the L4, and therefore could be found in limited numbers in those tractors described heretofore with L4 conversions. It had a DPA pump also.

The 6.288(TA) was derived from the P series and retained chain timing, although it was fitted with a DPA pump. It could deliver 65BHP @ 2000rpm.